Delta Science ContentReaders™

Heredity

Contents

Build Reading Skills

Preview the Book

You read nonfiction books like this one to learn about new ideas. Be sure to look through, or *preview*, the book before you start to read.

First, look at the title, front cover, and table of contents. What do you guess you will read about? Think about what you already know about heredity.

Next, look through the book page by page. Read the headings and the words in bold type. Look at the pictures and captions. Notice that each new part of the book starts with a big photograph. What other special features do you find in the book?

Headings, captions, and other features of nonfiction books are like road signs. They can help you find your way through new information. Now you are ready to read!

What Is Heredity?

MAKE A CONNECTION

Many animals can learn to do certain things. But animals also can do some things without having to learn how. Do you think storks must learn how to build a nest?

FIND OUT ABOUT

- why offspring look like their parents
- traits and heredity
- instincts
- the environment and traits
- learned behaviors

VOCABULARY

Heredity and Traits

What color are your eyes? Eye color is a feature, or **trait**. Living things get many traits from their parents. That is why children, or offspring, often look like their parents. The passing of traits from parents to offspring is called **heredity**. Traits that parents pass to their offspring are called **inherited traits**.

Inherited traits can be physical features, such as eye color and hair color in humans. People also inherit traits common to all humans, such as two eyes and a nose. In the same way, a zebra inherits a striped coat. A turtle inherits a hard shell. Plants inherit leaf shape and flower color.

Inherited traits also can be actions, or behaviors. Inherited behaviors are called **instincts**. Animals are born with certain instincts. For example, birds know by instinct how to build nests and care for their young.

✓ What is an inherited trait? Give an example.

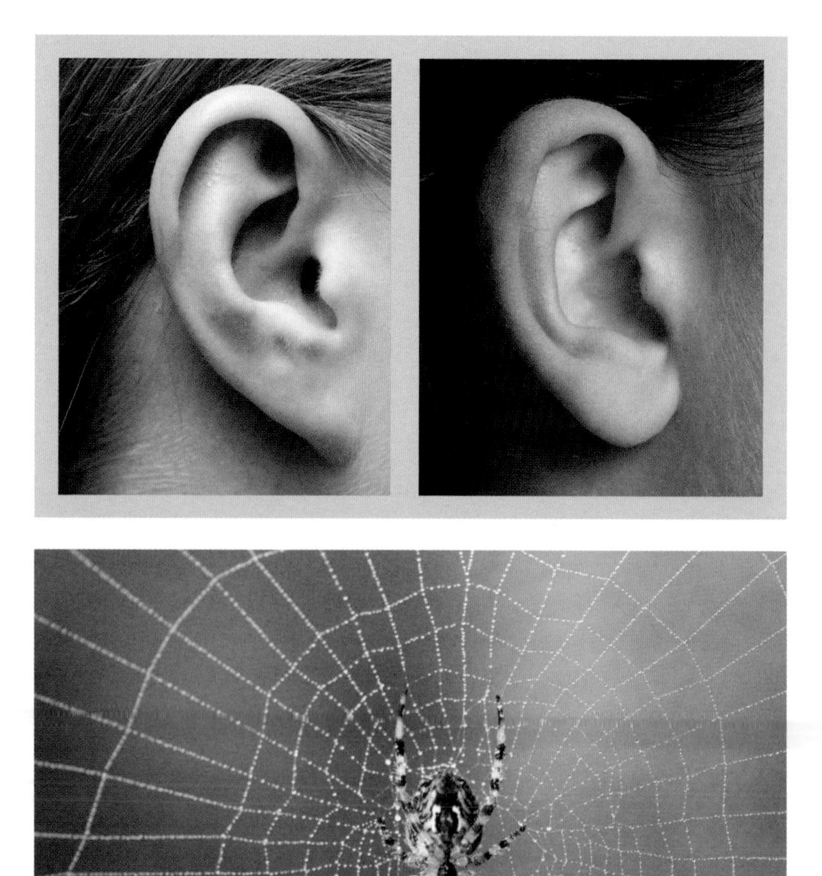

Earlobes can be attached or they can hang free. This is an example of an inherited trait. ▶

A spider does not have to learn how to build a web. Spiders are born with this instinct. ▶

A guide dog must be trained to lead a person safely across the street. This skill is a learned behavior.

The Environment and Traits

Living things, or organisms, also have traits that are *not* inherited. Instead, these traits are caused by the environment. A living thing's **environment** is all the physical things and conditions that surround it.

Physical traits that are not inherited are called *acquired traits*. For example, suppose you fell while running. Now you have a scar on your knee. The scar is a trait, but you were not born with it. It is the result of something that happened to you. It is an acquired trait.

Behaviors that are not inherited are known as **learned behaviors**. Unlike instincts, these behaviors are learned by watching others or from experience. Riding a bike is a learned behavior. So are tricks that a pet learns to do.

✓ You are reading this book. Is reading an instinct or a learned behavior? Tell why.

REFLECT ON READING

You looked at the words in bold type before reading. Choose a bold word on page 4 or 5 that you now understand better. Talk about the word with a partner.

APPLY SCIENCE CONCEPTS

Think about a family member or friend who looks like his or her parents. What are some traits that he or she inherited? Make a list of these traits.

Build Reading Skills

Sequence

Sequence is the order in which events or steps happen. Sequence is sometimes called time order or chronological order.

As you read pages 12 and 13, notice what steps take place when a cell divides in two.

TIPS

Thinking about the sequence of events can help you understand what you read.

- Ask yourself, "What happens first?" "What happens next?" and "What happens last?"
- Words such as *first, next, then, after, last,* and *finally* can be clues to the order of events or steps.
- Think about why the order of events is important. What would change if events happened in a different order?

A sequence chart can help you keep track of a sequence of events.

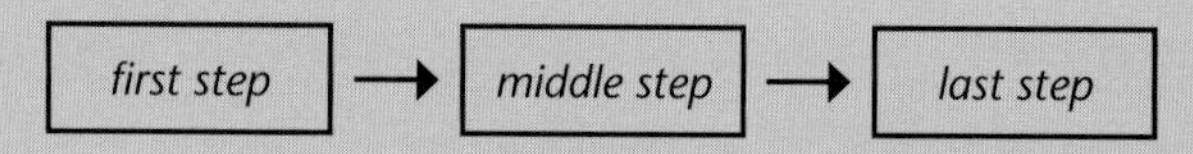

How Are Traits Inherited?

MAKE A CONNECTION

These baby swans are swimming with their mother. Why do you think a young animal grows up to look like its parents? Why doesn't it look like another kind of animal?

FIND OUT ABOUT

- cells
- chromosomes and genes
- two ways living things reproduce
- how and why cells divide

VOCABULARY

cell, p. 8
nucleus, p. 8
chromosome, p. 8
gene, p. 8
reproduce, p. 10

Genes

Cells are the tiny living building blocks that make up all organisms. Some organisms, such as bacteria, have only one cell. Other organisms, such as plants and animals, are made of many cells. The human body has trillions of cells.

The **nucleus** is the cell part that controls the activities of the cell. The nucleus also holds something like a set of instructions. An organism needs these instructions to live and grow. This set of instructions is in a special "code."

Chromosomes are cell parts in the nucleus that contain the code. Chromosomes are made of long strings of a material called *DNA*. DNA is divided into sections called genes.

A **gene** is the basic unit of heredity. Genes carry instructions for traits. All inherited traits are controlled by genes.

▲ A human is made of trillions of cells. Some simple organisms are made of only one cell.

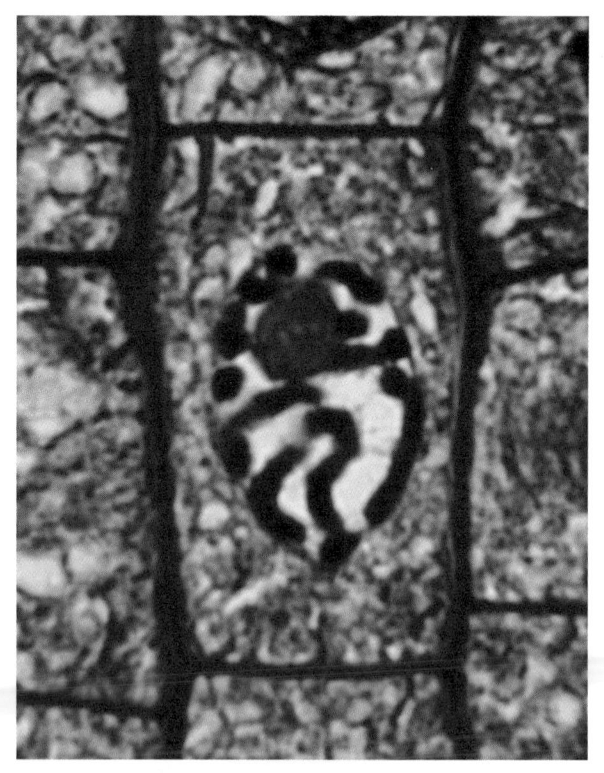

▲ The box shape is a plant cell. The large oval inside the cell is the nucleus. The dark lines in the nucleus are chromosomes.

One of a honeybee's genes controls many traits. The gene tells the bee when to search for food and what food to gather.

Each kind of living thing has a certain number of chromosomes in its cells. For example, a human cell has 46 chromosomes. A dog cell has 78 chromosomes. Each cell in an organism has the same chromosomes. Each cell has the same DNA and the same genes.

The DNA in a human cell has more than 20,000 genes. Each gene contains a single piece of information about a trait. Some traits, like dimples, are controlled by only one gene. Other traits, like hair color, are controlled by many genes.

Sometimes, one gene affects more than one trait. Honeybees have a certain gene that controls many traits. The gene tells the bee when to collect food for the hive. The same gene also tells the bee what kind of food to collect.

What is a gene?

Passing On Genes by Reproducing

Every kind of living thing makes more of its own kind, or **reproduces**. During reproduction, parents pass genes to their offspring. This is how a young organism inherits traits from its parents.

Reproducing With One Parent

Some kinds of living things can reproduce alone. Only one parent is involved. This is called *asexual reproduction*. In asexual reproduction, the offspring gets all its genes from one parent. So the offspring and parent are clones. This means that they have exactly the same genes. All one-celled organisms reproduce asexually. Some plants and animals can reproduce asexually, too.

A hydra is a tiny water animal. It reproduces asexually by a process called budding. Its body grows a bump, or bud. The bud grows into a new hydra and then breaks off.

Some sea stars reproduce asexually. They may split in two or lose an arm. Even one arm can grow into a new sea star. This kind of asexual reproduction is called regeneration.

Hydras can reproduce by budding, a kind of asexual reproduction. ▼

A sea star can reproduce by regeneration. One arm can grow into a new sea star. ▼

Kangaroos and most other animals make more of their own kind through sexual reproduction. The offspring receives genes from both its parents.

Reproducing With Two Parents

Many kinds of organisms need two parents to reproduce. Reproduction with two parents is called *sexual reproduction*. In sexual reproduction, special cells from each parent join together. These cells are known as sex cells. Female sex cells are called eggs. Male sex cells are called sperm. When an egg and a sperm join, a new cell called a fertilized egg forms. This cell can develop into a new living thing. Almost all animals come from fertilized eggs.

A sex cell has only one set of a parent's genes. All the other cells in the body have two sets of genes. When sex cells join, the fertilized egg has one set of genes from each parent. So it has a full two sets of genes.

✓ Tell how asexual reproduction and sexual reproduction are different.

Cell Division

Living things made of many cells actually begin life as just one cell. They grow and develop by adding new cells. Cells are added by a process called cell division. Cell division also is the way that one-celled living things reproduce.

In cell division, one cell divides, making two cells. Then those two cells divide, making four cells. Cells keep dividing until a living thing has all of its cells.

Cell division continues throughout life. This is how an organism's body grows and repairs itself. For example, you grow taller and your hair grows longer because of cell division. Also, old cells wear out. So cells divide, making new cells to take their place.

Cell division happens in a special order, or sequence.

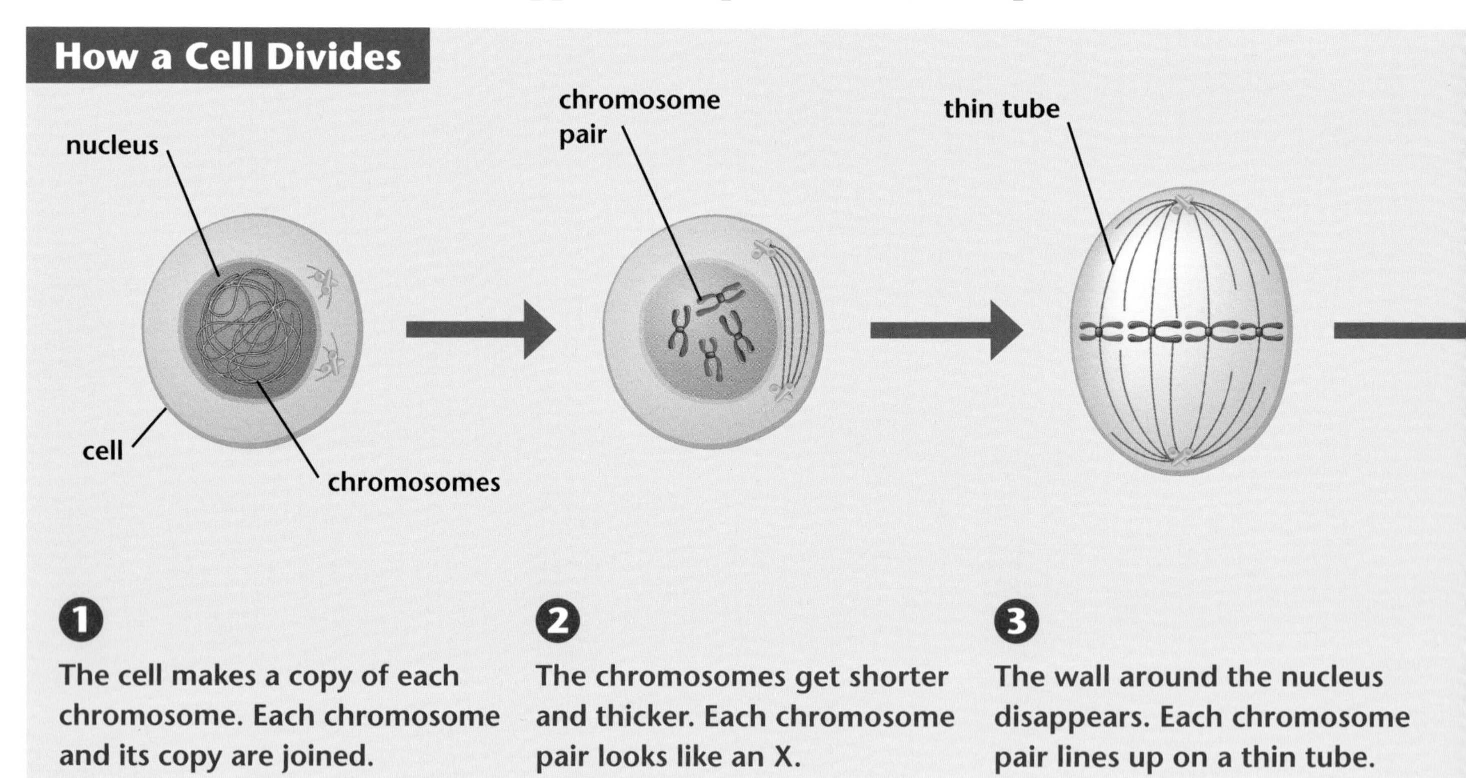

▲ When a cell divides, it makes two cells that are the same as the first cell.

First, the cell makes a copy of each chromosome. Each chromosome and its copy are joined.

Next, the chromosomes get shorter and thicker. Each chromosome pair looks like an X.

Then, the wall around the nucleus disappears. Each chromosome pair lines up on a thin tube.

Soon, each chromosome pair comes apart. The parts move away from each other.

After that, the cell makes a new nucleus around each set of chromosomes.

Finally, the cell pinches together and divides in two. The two new cells have the same number of chromosomes as the first cell. They have the very same DNA. They are exact copies of the first cell.

Why do cells divide?

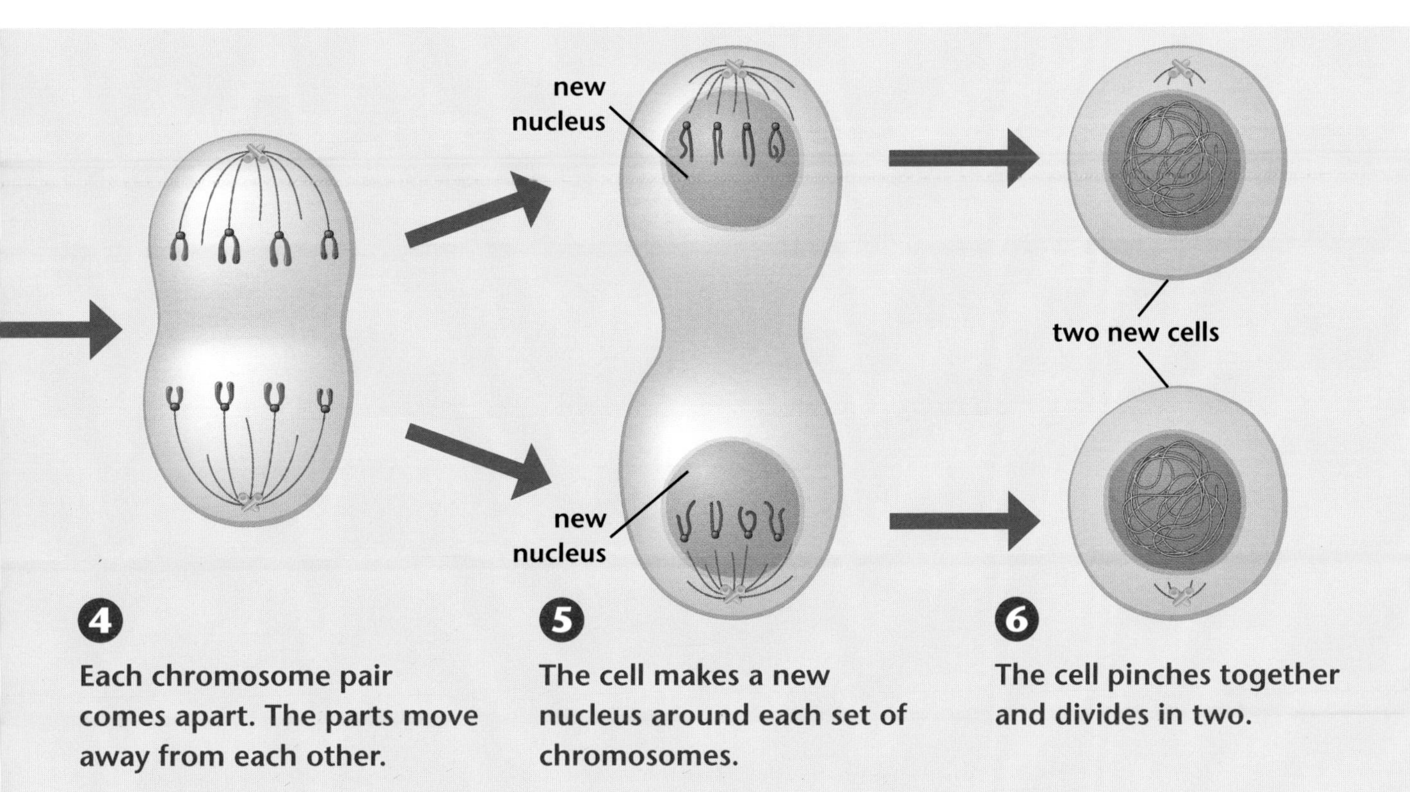

4 Each chromosome pair comes apart. The parts move away from each other.

5 The cell makes a new nucleus around each set of chromosomes.

6 The cell pinches together and divides in two.

REFLECT ON READING

Work with a group to make a set of cards showing each step in cell division. Do not include the step numbers. Take turns putting the cards in the correct order, or sequence.

APPLY SCIENCE CONCEPTS

Genes give instructions to your body. Make a list of some things that genes tell your body to do. Compare your list with a partner's.

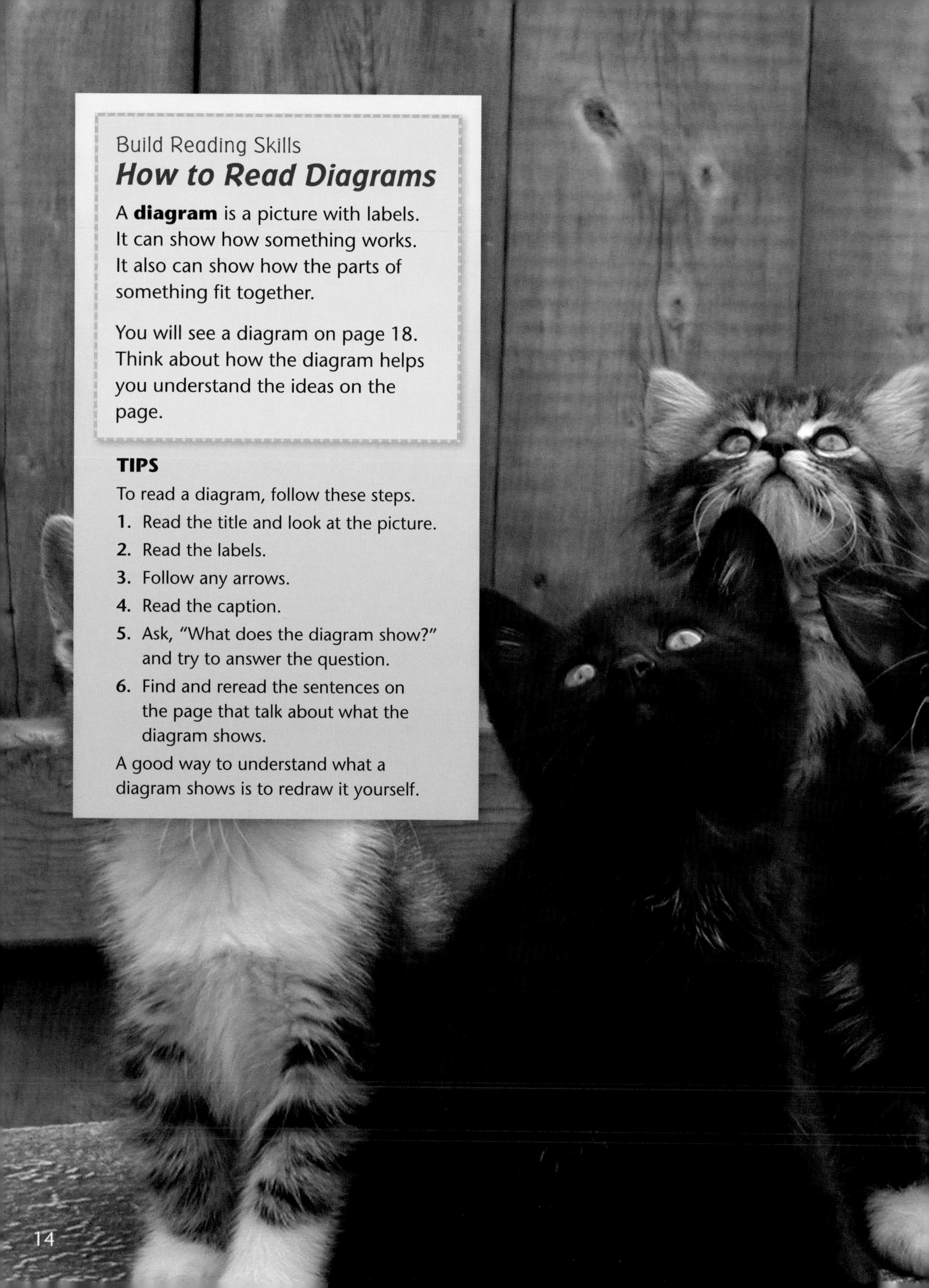

Build Reading Skills

How to Read Diagrams

A **diagram** is a picture with labels. It can show how something works. It also can show how the parts of something fit together.

You will see a diagram on page 18. Think about how the diagram helps you understand the ideas on the page.

TIPS

To read a diagram, follow these steps.

1. Read the title and look at the picture.
2. Read the labels.
3. Follow any arrows.
4. Read the caption.
5. Ask, "What does the diagram show?" and try to answer the question.
6. Find and reread the sentences on the page that talk about what the diagram shows.

A good way to understand what a diagram shows is to redraw it yourself.

What Are Variations?

MAKE A CONNECTION

These kittens all have the same parents. Why don't they look exactly the same? What differences do you see?

FIND OUT ABOUT

- differences among individuals and how they happen
- Mendel's experiments with pea plants
- dominant and recessive traits
- crossbreeding and hybrids

VOCABULARY

variation, p. 16
dominant trait, p. 18
recessive trait, p. 18

In Mendel's first cross, the purple-flowered parent had two "purple" genes. The white-flowered parent had two "white" genes. Each parent gave one gene to the offspring. The chart below shows this cross.

Each offspring received one purple gene and one white gene. But all the offspring had purple flowers. Why? Having purple flowers is a strong trait, or **dominant trait**. Having white flowers is a weak trait, or **recessive trait**.

An organism with two dominant genes for a trait has the dominant trait. An organism with two recessive genes for a trait has the recessive trait. What if an organism has one of each gene? Then the dominant trait "hides" the recessive trait. The dominant trait shows. In humans, having dimples is a dominant trait. Having blue eyes and being tall are recessive traits.

✔ Why did Mendel study pea plants?

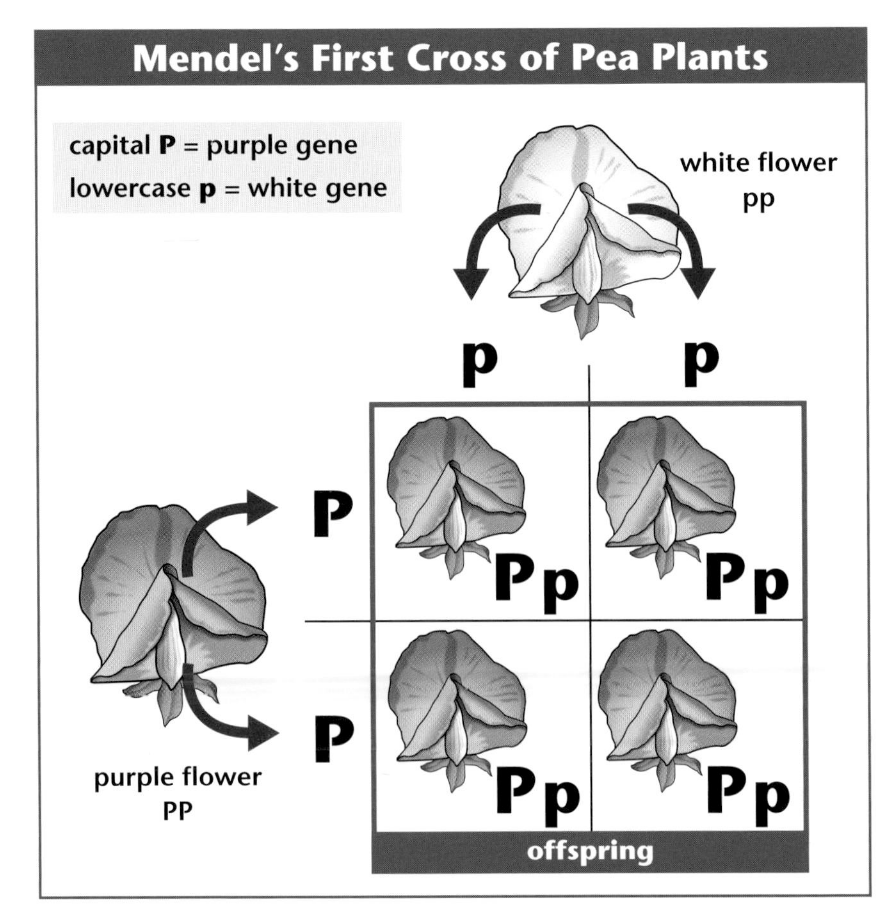

The diagram shows a purple plant crossed with a white plant. All the offspring are purple. Purple flowers are dominant in pea plants. ▶

Crossbreeding and Hybrids

A *species* is a type of living thing. A species can have many varieties. Crossing two varieties of one species is called *crossbreeding*. Mendel crossbred pea plants. Farmers crossbreed plants such as corn to get crops with certain traits. People have crossbred dogs. Hundreds of breeds, from Chihuahuas to Great Danes, are the result. Crossbreeding also can happen in nature.

Sometimes, people cross two different species. The offspring of this cross is called a *hybrid*. A hybrid has traits of both species. A mule is a hybrid of a female horse and a male donkey. Hybrids usually cannot reproduce. Also, hybrids do not usually happen in nature.

✓ What is a hybrid?

A mule is the offspring of two species, a donkey and a horse. A mule is a hybrid. ▶

REFLECT ON READING

Look again at the diagram on page 18. Tell what the diagram shows in your own words.

APPLY SCIENCE CONCEPTS

Suppose you want to make a better plant. In your science notebook, list the plants you will cross. They can be from the same or different species. What traits should their offspring have?

Build Reading Skills

Main Idea and Details

The **main idea** of a paragraph or part of a book is the most important point. **Details** give more information about the main idea.

As you read page 22, look for the main idea about adaptations.

TIPS

The topic sentence tells the main idea of a paragraph. It is often the first sentence in the paragraph. To find the main idea, ask, "What is this paragraph mostly about?"

Details may answer Who, What, When, Where, Why, and How questions about the main idea. Details can be

- examples
- descriptions
- reasons
- other facts

A concept web can help you keep track of the main idea and details.

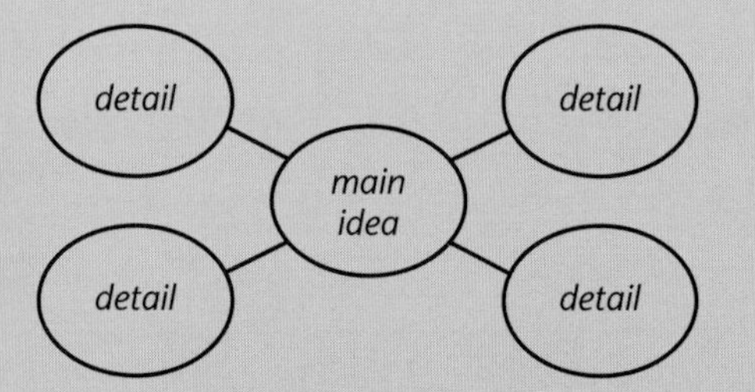

How Do Traits Help Organisms Survive?

MAKE A CONNECTION

Camels can live in the desert, but many other kinds of animals cannot. Why do you think camels are able to live in this hot, dry environment?

FIND OUT ABOUT

- adaptations to the environment
- how species change over time

VOCABULARY

adaptation, p. 22
diversity, p. 23

Adaptations

Inherited traits that help a living thing survive in its environment are called **adaptations**. An adaptation might help a plant get water or make seeds. An adaptation might help an animal get food, find shelter, reproduce, or protect itself.

Some adaptations are physical parts. Beaks, claws, feathers, and wings are physical adaptations. A bird's beak helps the bird get and eat certain foods in its environment. An elephant's trunk helps the elephant pick leaves and drink water. Plants also have physical adaptations, such as thorns that help keep animals away.

Other adaptations are inherited behaviors, or instincts. Migration is an example. Some birds fly long distances, or migrate, at certain times of the year. Birds may migrate to an area where they can more easily find food in winter.

✓ What is an adaptation? Name two adaptations of birds.

▲ The sharp, curved beak of an eagle is a physical adaptation. The beak helps the bird catch, hold, and eat its food.

▲ Some adaptations, such as migration, are behaviors. Snow geese nest in the Arctic and then migrate south to find food.

This drawing shows how horses probably looked millions of years ago. Over time, the species changed.

How Species Change

All the living things of the same kind belong to the same species. Individuals in a species have a variety, or **diversity**, of traits. Some traits are especially helpful for survival. Examples are an eagle's curved beak and an elephant's long trunk.

Individuals with helpful traits are more likely to survive. When they reproduce, they pass on genes for the helpful trait to their offspring. As a result, all the individuals in the species eventually inherit the helpful trait. This is how species change over time.

Most changes in species happen very, very slowly over many generations. Early horses had short legs. But individual horses with longer legs could run faster. They could get away from danger. These horses survived to pass on their genes. Over millions of years, because of heredity, horses came to have long legs.

Tell how a species can change over time.

REFLECT ON READING

Make a concept web like the one on page 20. Use the web to organize what you learned about adaptations. Put the main idea in the middle. Add details, such as examples and other facts.

APPLY SCIENCE CONCEPTS

Look at an animal near your home or on TV. Notice one of its body parts or behaviors. Think about how this adaptation helps the animal live. Talk about your ideas with a partner.

Glossary

adaptation (ad-ap-TAY-shuhn) an inherited body part or behavior that helps a living thing stay alive in its environment **(22)**

cell (SEL) the smallest unit of a living thing; a building block of life **(8)**

chromosome (KROH-muh-sohm) a cell part made of a material called DNA; it carries the genes that give a living thing its traits **(8)**

diversity (di-VUR-si-tee) variety **(23)**

dominant trait (DOM-uh-nuhnt TRAYT) a trait that appears if a living thing has one or two genes for that trait **(18)**

environment (en-VYE-ruhn-muhnt) all the physical things and conditions, such as soil, air, climate, plants, and animals, that surround a living thing **(5)**

gene (JEEN) a section of DNA on a chromosome that has information for a trait **(8)**

heredity (huh-RED-i-tee) the passing of traits from parents to offspring **(4)**

inherited trait (in-HER-i-ted TRAYT) a feature passed from parents to offspring through genes **(4)**

instinct (IN-stinkt) an action that an animal knows how to do without being taught; examples include finding food, caring for young, and building a shelter **(4)**

learned behavior (LURND bi-HAYV-yur) an action that an animal learns by watching others or from experience **(5)**

nucleus (NOO-klee-uhs) the part of a cell that controls the cell's activities **(8)**

recessive trait (ri-SES-iv TRAYT) a trait that appears only if a living thing has two genes for the trait **(18)**

reproduce (ree-pruh-DOOS) to make more of one's own kind **(10)**

trait (TRAYT) a feature, such as a body part or behavior, of a living thing **(4)**

variation (vair-ee-AY-shuhn) a difference between two or more individuals in a species **(16)**